Unus pro omnibus, omnes pro uno

Team of 5 Million

Kathiravelu Ganeshan

@Ganeshan

Preface

New Zealand is a beautiful country, an enormous playground for people of all ages, with an interesting and unique Kiwi way of thinking. For example, when just four people out of nearly 30,000 escaped from quarantine, posing significant Covid-19 scale danger to the entire population of 5 million people the government did not stop at charging the four who broke the law; they started providing even more support to the people in quarantine.

New Zealanders also punch way above their weight in many areas, especially in sports. I do not believe any other nation has so many people performing at top international level in so many sports, at least on a per capita basis.

The entire population, including those who genuinely believed that it was not possible to rid the country of Covid-19, stood together as one team and stopped the spread of the virus.

Within a few weeks, we were able to, once again, sit close together and watch rugby matches in packed stadiums, go to school, work, and carry on our normal lives without having to wear masks, practise social distancing, or use hand sanitisers.

We achieved this without sacrificing the lives of our nanas, grandpas, and of those with diabetes, obesity, asthma and other respiratory disorders, cardio-vascular diseases and immune deficiencies, at the altar of economy.

We knew that we can recover the economy, and that we cannot bring the dead back to life.

Now that we have broken free from the grip of Covid-19, it is time to address of issue of economy, thinking differently, embracing our famous number-8-wire mentality, and working together as a Team of 5 Million, just like we did, to beat Covid-19.

In this mini-book, I suggest that instead of looking at Covid-19 as a problem, which indeed it is, we make use of the opportunities Covid-19 has created.

Together, we can dream big and achieve a robust economy and an improved quality of life for people at every stage of their life.

New Zealand can be the All Blacks of economy and quality of life.

We cannot do this if let ourselves get distracted by petty politics, naysayers, those who still believe that the earth is flat, and ants-in-the-pants commentators.

Patience and a clear long-term vision are essential for success, not lies and negativity. Let's do this.

The World is changing faster than ever before

As at 14 July, 2020, the world is still in the early stages of a global pandemic, not in the middle.

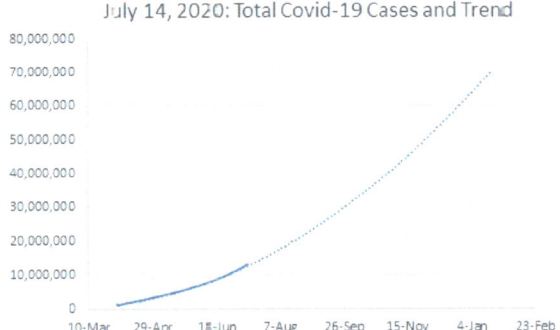

It has been reported that there is at least another virus that is not that far from starting a second pandemic.

On top of viruses, we have more and more of other significant destructive events happening; increasing pollution, rising temperatures, melting glaciers, rising sea levels, coastal erosion, tsunamis, and, of course, tornadoes and cyclones.

New Zealand also sits on top of the major fault line between the Pacific and Australian tectonic plates, and is therefore subject to as many as 15,000 earthquakes per year (only 100-150 are large enough to be felt), and the occasional volcanic eruptions.

Just over a 116 years ago, Wright brothers flew a few meters with their first powered aircraft.

In less than a hundred years, we had over half a million people in the air at any time. In one fell swoop, the coronavirus grounded most flights, and brought the airline industry to its knees. Many airlines have been forced to seek government support to ride out the epidemic.

Cruise liners that were larger than hotels became floating death traps for many people, with countries that previously welcomed these floating hotels with much fanfare now refusing entry to these same vessels.

The virus used aircraft and cruise ships to reach every corner of the world.

Steam engines which powered locomotives for about 200 years were replaced by diesel engines, which in turn had to make way for electric trains. Even some of the tracks have been superseded by magnetic levitation.

Driverless cars are already on our roads.

Space tourism is on the cards, and we are even contemplating settling humans on Mars.

We can access vast amounts of data, information and knowledge with touches and swipes of a screen, or voice commands.

We have machines that think, learn, and even advise us on so many things from spelling and grammar to healthcare.

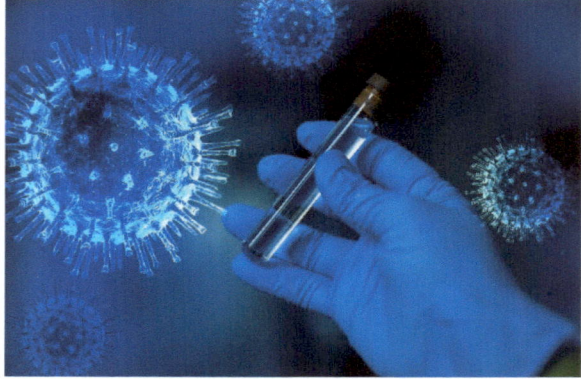

Yet, we are unable to develop a vaccine that will protect us from the virus, or find a cure for the disease caused by the virus. All we can do is to try and hide from it.

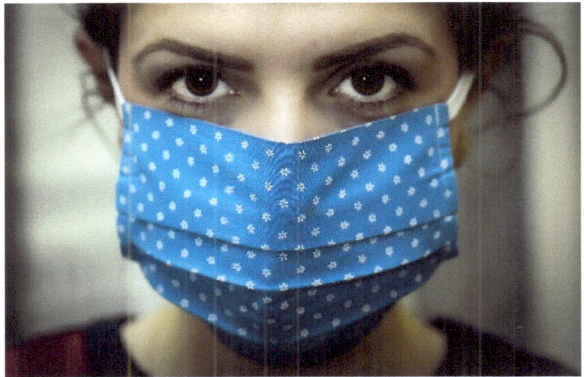

Managing the pandemic

Our greed, our way of life, and our dependency on so many physical *necessities* has turned us, humans, into a species that is relentlessly working and accelerating towards self-destruction as well as the destruction of the planet itself. We have already made many species of plants and animals extinct.

Even in the face of a major global pandemic, we are not willing to cooperate as a species to stop the spread of the virus.

Only a few countries had the vision and determination to contain the virus. The solution was, and still is, deceivingly simple. All they had to do was to deny the coronavirus its transmission pathways. These countries can still control the spread by simply blocking the route of transmission.

Taiwan, even though it was geographically very close to the initial Covid-19 hotspot, managed to contain the virus in a very smart way. The government and the general population were well prepared in terms of the physical resources and the community spirit needed to tackle a dangerous virus like this one; they instantly recognised, and attacked the organism's Achilles' heel, its transmission path. They learned from their prior experience with SARS in 2003, and used that knowledge extremely well; they did not hesitate. Instead, they acted promptly and decisively.

New Zealand is lucky in many ways. It is tucked away in a corner, far from the rest of the world. Except for its main cities, it is sparsely populated. Our health experts are on top of their game. The government listened to these health experts.

The government was also; brave enough to take a bold decision; quick enough to act; and, communicative and persuasive enough to get the cooperation of the 5 million, the entire population of this small nation.

The population was smart enough to; understand the seriousness of the problem; accept the costly but effective solution proposed and implemented by the government; ignore the noise from the few loud and vocal naysayers, publicity seekers, armchair critics and even pressure from some *world-class, renowned* epidemiologists; and, act responsibly.

For a number of reasons, other countries, chose to follow different paths and are continuing to pay the price in terms of lives and livelihoods. Many are still walking around with masks and/or social and physical distancing and living in fear of catching the virus.

Some are just about to surrender to the virus, opting to *live with it.*

It is indeed, ironic that humans who have wiped out so many species for good, and have enough firepower to destroy all remaining species, including their own species, on the planet at the touch of a button have no choice but to surrender to Covid-19 and opt to share the planet with the enemy, a half-life!

Epidemiologists, economists and others have been studying and will continue to study, for years to come, the coronavirus, the responses of various countries and the results.

It is my view that studying, in detail, the responses and results (in alphabetical order) of Australia, Czechia, Germany, Hong Kong, New Zealand, Sri Lanka, Sweden, Taiwan, UK, USA and Vietnam will help us greatly in the future when another similar virus starts to spread.

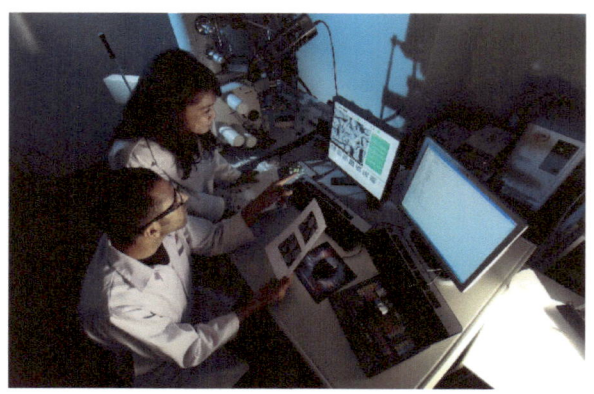

Globalisation: Shortcomings exposed

One of the many things that the pandemic exposed is how the beautiful utopian ideas of a global village and a global economy do not work during bad times. While I am not a fan of Donald Trump, I agree with him in that countries need to be self-sufficient.

Global trade and the idea of a global village have resulted in countries becoming dependent on other countries for food, and several other basic essential necessities, even life-saving supplies.

We export and import bottled water. Even canned fresh air has found a market in certain cities with extreme levels of air pollution.

The pandemic created a huge global demand for essential items like masks, ventilators, test kits, and certain medicines. We only need to look at how some countries, and even states within countries behaved during the pandemic to understand why going global has serious shortcomings.

Doctors, all of whom took the oath to save lives, had to make calls on who lives and who dies. Patients were denied life-saving treatment because there were not enough ventilators. Hospitals in some countries even refused admission, while others set up entire hospitals within days Many doctors and healthcare workers themselves died due to lack of supplies of Personal Protection Equipment (PPE).

Most countries have the knowhow, technologies and the resources, including human resources, to produce many of these essential goods.

Thinking about the future of New Zealand, I suggest that we develop industries that produce, or at least are capable of producing at short notice, most of the essential supplies in case of a second pandemic, or other major destructive event. Such industries will create jobs and increase our GDP with products that matter. We have the technologies and the people to develop such industries, create jobs, earn export income and, most importantly, make New Zealand self-sufficient, in essential products.

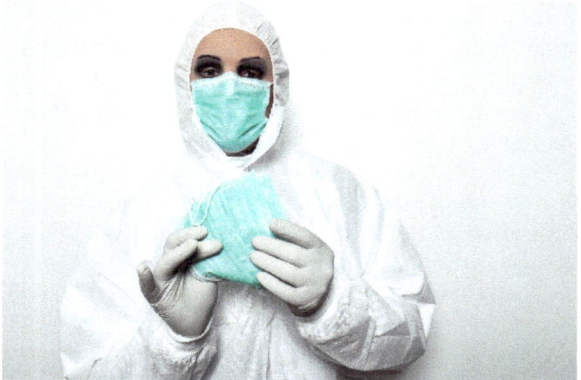

Export of goods

In 2019, New Zealand shipped $38.2 billion worth of goods to overseas countries. This figure, when divided by the population of 5 million people works out to $7600 per year, per every adult and child in New Zealand.

Asian countries received 58.3% of New Zealand's exports, while 15.5% were sold to other countries in the Oceania region with Australia taking the bulk of this 15.5%. North America's imports were 11.2% with Europe taking the next place with 9.9%, followed by African countries at 2.6%.

In 2019, New Zealand's top 10 exports accounted for about 75% of the overall value of its global shipments.

These products included mostly food products such as dairy, eggs, honey, meat, and fish. Wine and wood were the other two major contributors.

I do not see any reason why these exports are, or will be, affected by Covid-19.

However, image is an important factor that influences consumers.

Keeping New Zealand Covid-free may help enhance our image as a country whose food products are safer to consume, just as our clean, green image adds value in so many ways.

Tourism

During the 12 months to March 2019, total tourism expenditure was $40.9 billion. Another $11.2 billion was contributed indirectly by industries that support our tourism sector.

International tourism, our biggest export earner, contributed $17.2 billion. This is 21% of New Zealand's total exports of goods and services.

The tourism industry directly employed 229,566 people. This is 8.4 percent of the total number of people employed in New Zealand.

Many of the operators in the tourism business have made significant investments and may not survive the fallout from Covid-19.

It is understandable that they want the borders open as soon as possible. Some of them believe that tourists will start coming in and that business will revert to pre-Covid-19 levels.

The reality is that Covid-19 is just beginning its destruction. It will last at least a couple of years.

Will tourists really start coming in if New Zealand is not a Covid-free country? Will they be prepared to sit close to each other in transport and attractions? Will they be willing to touch and hold the handles, handrails and other physical objects in the tourist attractions?

In the case of Asian tourists, who form a major part of the market, will they be willing to accept that New Zealanders are not wearing masks in public?

Taking into account the above ground realities, I believe that it is in the best interests of the tourism and related businesses to do everything they can to keep New Zealand Covid-free.

Around 1.18 million Australians visit New Zealand each year. Australia is our biggest tourism market. Unfortunately, parts of Australia are finding it difficult to contain the virus. A trans-Tasman bubble seems to be an elusive goal at least for the next six months.

China, our second largest tourism market with nearly a quarter of a million visitors, is also struggling to contain the virus.

There is still however, an option for tourism businesses to consider. They can shift their focus to local tourism and this has already started happening.

Given the Covid-19 situation around the world, not many New Zealanders would risk travelling outside New Zealand in the next few months.

It is a great opportunity for New Zealanders to enjoy their own country, and do a lot more with the money they save on airfares, among other things.

In 2017, Kiwis took 2.9 million trips overseas. Taking the average cost of a return ticket at $1000 that is just under $3 billion. It is fair to assume that the 2020 projections would have been higher. Add to this all the other funds Kiwis would have spent overseas – likely to add up to well over $6 billion per year. While this is only about a third of what international tourism usually brings in, it may still keep some operators going while the world grapples with Covid-19.

Taking the above into account, I believe that it is in the best interests of our tourism operators to keep New Zealand Covid-free.

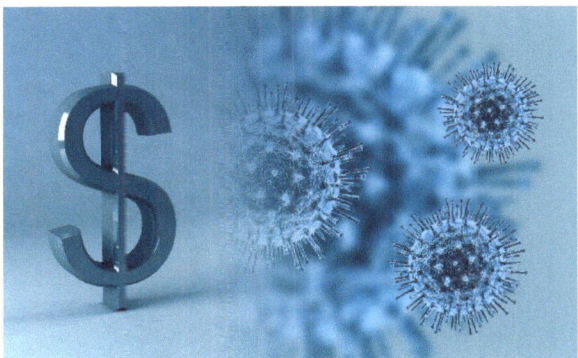

International students

Universities, polytechnics, private colleges and even schools depend to some degree, or entirely, on international students for income.

Covid-19 has struck a severe blow to all these institutions; some may not survive. Let's not forget that some, including the biggest polytechnic, were struggling financially even before Covid-19 struck and had to be rescued by the government.

Some of the universities also had several rounds of redundancies and other cost-cutting measures over the past two decades.

International students contribute around $5 billion a year to the economy. This income translates to $1000 per year to every child and adult in New Zealand.

Covid-19 has given us the opportunity to double, even quadruple, this income. We have more than enough well-qualified educators with adequate qualifications, training and experience in every discipline, and adequate physical infrastructure to handle much larger numbers.

From the perspective of international students, and their parents and families where appropriate, a Covid-free country will always be a safer, and therefore preferred, choice than a country that is unable to contain the virus. A 14-day quarantine is a relatively minor inconvenience, compared with living for three or more years as a student wearing masks, washing hands often and maintaining social and physical distances.

The issue of international students, once again, needs significant clever thinking, conversations with all the stakeholders and time to come up with a sound plan.

We cannot trust universities to come up with a sound plan, or to manage any isolation facilities; that is not their core business; they have no experience doing this; and, there will be conflict of interest.

On the other hand, a government-controlled facility that will take in all arriving students and keep them in isolation for 14 days in one or more locations, away from our cities and other population centres would work well.

An arrangement in which international students can come in as a continuous stream, rather than in two lots in February and July, will make it easier for the government to maintain a year-round isolation facility staffed by adequately trained people, and thus ensure that New Zealand remains Covid-free.

Universities and polytechnics will also benefit by getting a much higher utilisation rates for their real estate and other infrastructure. They need to come to the party, play their part and plan for a continuous intake of students, abandoning their two-semester model – a model that should have been ditched decades ago.

Universities, polytechnics and other education providers also need to move away from teaching, to facilitating learning. I have dealt with this topic in some of my previous mini-books – available on Amazon.

Due to the nature of educational institutions, the number of students and staff involved, how close people get to each other physically in classrooms and laboratories, the long hours in these confined spaces and the number of interactions each student has with others each day, just one new Covid-19 cluster in such institutions, would have already reached community transmission levels, by the time the first case shows symptoms.

Contact tracing may be near impossible, compounded by the fact that most of our universities, polytechnics and private colleges are located in our crowded cities.

Therefore, it is in the best interests of universities, polytechnics and other education providers to keep New Zealand Covid-free.

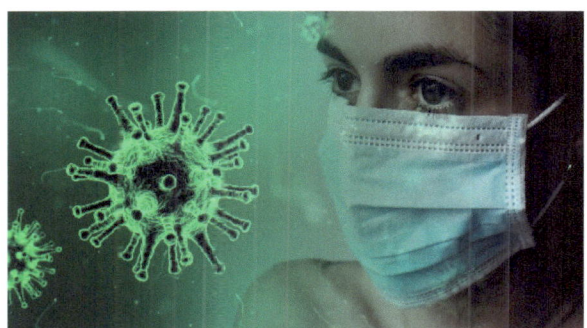

Youth unemployment

A couple of years ago, during Christmas break, my family and I visited Whangarei for a few days. That was the first time we stayed in Whangarei and visited several tourist attractions in the city and surrounds.

Carparks in every one of the tourist attractions we visited had one or more volunteers watching over the cars parked there. Large signs urged visitors not to leave valuables in their cars.

On the last day of our trip, we visited the A H Reed Memorial Kauri Park. There was a volunteer security guard watching over cars in the main car park. We could find parking for only one of our two cars in the main carpark. We asked the volunteer security guard if it was safe for us to park our other car in the overflow carpark, less than ten meters away from the main carpark, and he told us that it would be fine. There were at least another dozen cars parked in the overflow carpark.

We were inside the park for about half an hour. Just as we were exiting the Kauri Park, we heard some noise, cars driving off and people running towards the overflow carpark. We rushed there to find our rear windscreen smashed in at least two places with a hammer or something similar.

Initially, of course, I was very angry. But as I reflected on the event, I started thinking that we can, and should, do more to help people like the ones who broke my car's rear windscreen.

It is my belief that most people engage in destructive behaviours, more often than not, because they fall through the cracks in our systems, especially our education system.

During my 45 years in education, I helped many *lost souls* to find their way in the real world, with some now earning much more than I do.

Education is the key to many of the problems that we face in each community, state and country around the world.

In 2020, even during this current global coronavirus pandemic, the New Zealand government invested heavily in our schools, which is commendable.

There is more to be done. Schools need to move away from their current models of teaching and theory examinations, and shift towards real education. They need to find better ways of motivating learners.

They need to consider a *Sister Act* mode of engaging learners.

Several schools around the world, including a few in New Zealand have begun this major transformation.

One of the reasons for me to write this book is to encourage other schools, to take a look at the select few that have started this transformation in education and see how they too can join in and help transform learning into a much more meaningful and enjoyable activity. I have dealt with this topic in some of my previous mini-books – available on Amazon.

Educational institutions, intermediate and secondary schools in particular, need to stop people falling through the cracks. This will prevent young people starting to believe that hurting others, destroying property and even suicide as solutions.

Suicides

Several countries in the world are grappling with the issue of suicide, especially among their youth and the elderly.

During the lock down in New Zealand, a small number of people argued against the lock down claiming that there will be more suicides. However, it turned out that the lock down did not result in an increase in the suicide rate; perhaps, because of the minimal length of the lock down compared with some other countries; perhaps, because Dr Ashley Bloomfield, the Director-General of Health and Ms Jacinda Ardern, the Prime Minister, fronted up daily at 1 pm and gave briefings which most people found to be very informative and reassuring.

What we saw was a clear plan that was continuously tweaked as more data and knowledge on how the virus spread became available. We saw two unflappable, knowledgeable leaders in sync with each other, armed with the latest information and showing tremendous empathy and a lot of patience with the journalists, even with those who asked the same questions again and again and the ones who asked some loaded questions.

Imagine having the patience to answer, questions about 5G causing Covid-19 and asking the Director-General of Health to demonstrate how to wear a mask properly. Many viewers found such questions and the answers quite entertaining. I believe that the levels of ignorance and arrogance that a few journalists demonstrated probably made some people who considered committing suicide feel better about themselves – just a tongue-in-cheek hypothesis that needs further study.

Coming back to the issue of suicides in the community in general, two particular groups need more attention; youth and seniors. I will deal with the physical and mental health of seniors in a later chapter.

Youth suicides, in my view, can be reduced if our education systems improve. Our education systems are antiquated. Unlike physical antiques some of which can be sold for a lot of money, our antiquated education systems are doing a lot of damage to humanity and are also a contributing factor to youth suicides.

Education should be holistic and include learning about physical and mental health and wellbeing – not just science, arts and sex. Physical and mental health education should be an important, compulsory part of the education at all levels from 1 to 10. To put things in perspective, consider what percentage of people successfully complete their masters and doctorates, and the mental health of some of these people during studies and even after completion. That is only one half of the story.

The second half is that there are not just cracks, but huge holes in our education system. If we take a good look at what we do, we will find that we can do a lot more to shift the focus to meaningful, enjoyable, active learning – like some schools and tertiary institutions are already doing.

By making learning an enjoyable, achievable activity, we can increase student engagement and reduce youth suicides.

Written theory exams at every level is a cause of stress for students of all ages, their parents and even teachers and lecturers. I see no reason why we are still using so many written theory examinations. One of my classmates in school committed suicide at age 14 as he did not do well in his examination. We can find lots of such stories in many parts of the world.

Written theory examinations are not the best measures of student learning, either.

You may also want to check out this 1 minute video.

https://twitter.com/jeremycorbyn/status/1118398182629629953?lang=en

High-tech industries

We already have some great hi-tech industries in New Zealand. This includes Rocket lab, Weta Digital and a number of Games Developers. We also have some world-beating film production facilities. And, there are plenty more.

To develop such highly profitable, high-profile, high-tech industries, we need people, people with the right knowledge, skills, attitude and passion.

We can continue to, and have to, import more skilled people from other countries. We can also help develop local talent, talent that is otherwise wasted. I deal with this in the next chapter.

Putting more effort into developing local talent

One of my former colleagues told me that if all the *Indians* working in IT in New Zealand decided to leave, many companies will be in serious trouble. It is hard to imagine all the *Indians* leaving New Zealand *en masse*. On the other hand, if we had a major problem with a pandemic like Covid that results in massive deaths in New Zealand, people may abandon New Zealand – who knows.

New Zealand IT sector of course has people from all over the world, not just from India. Many other sectors also have people like doctors, nurses and seasonal workers from all over the world. All this adds to the diversity, brings in talented people at various levels and makes the country richer – economically and culturally.

However, we also need to spend more time and effort into developing more local people to work in some of these high-tech and high-need areas.

Robotics, AI, machine learning, data science, internet-of-things, games and App development, virtual reality, telemedicine, farm automation, driverless vehicles, and many other areas are going to be the money-makers in the near future, if they are not already.

Trying to make money on the property market, selling coffee, beer and food will all keep the economy going but not for long. For example, as soon as we came out of lock down, some city councillors and politicians were encouraging people to go back to their city offices so the coffee shops, bars, takeaways and other businesses can have more business and the building owners in the city can collect rents and the busses and trains will collect fares.

Things are not going to go back to the pre-Covid *normal* because Covid-19 has helped us understand that there are better ways of living, making a living, keeping the economy going and running a business and, at the same time, reduce pollution, suffering and stress. For example a major New Zealand bank is closing several branches – they have found that customers prefer to do things online and employees are happier and more productive when they work from home. The commercial property market is going to take a hit as more businesses follow suit.

On 12 June 2020, I noticed a petrol station cafe and shop permanently closed – the pumps are still open and customers pay at the pump. At least a few people lost their jobs for good. However, the new EFTPOS machines and the related hardware and software as well as the electronic security systems need maintenance – so the need for tech employees.

This is an example of just one of the areas in which we need to inspire and engage local students.

The world is waking up to telecommuting. Many high-tech companies now have most of their employees working from home. In 1984, when I suggested, during my conference presentation to a group of medical doctors that one day we will have telemedicine, they did not believe me. Telemedicine and even tele-surgery are now happening. Robotic and robot-assisted precision surgeries are already happening.

Covid-19 necessitated that we stockpile masks, ventilators and other items. We had problems securing enough of these.

What all this means is that we need more hi-tech industries and people.

Thinking back to my broken car windscreen, I am of the view that we can, and should, find ways of inspiring and educating our young people in these areas, instead of producing graduates in disciplines that do not have adequate job opportunities with the graduates often ending up flipping burgers or peddling something or other.

Changing the education paradigm, making learning more meaningful and enjoyable, minimising theory examinations, and thus inspiring and engaging more students in science and technology will result in a win-win for all of us. It will also result in less youth suicides.

Senior citizens

As an elderly person myself, I find the discrimination quite staggering. Sometimes it is positive – like the time in a crowded bus several students from one secondary school offered me their seats. At other times, it is quite negative.

I am lucky to be in New Zealand and, along with the team of 5 million, I am thankful that we have quashed the virus and are doing our best to keep it out.

I am also in great health and will most likely survive Covid-19 if I am unfortunate enough to contract it.

It is indeed, quite shocking that in certain countries during the pandemic the elderly were denied the care they deserved for a number of reasons.

I had a Twitter account since April 2009. It was dormant for many years. But when I was in lock down I started using it again and even became an addict of sorts.

The kind of tweets that were flying around indicated that there was a minority of people who were ready to sacrifice their grandparents and parents at the altar of economy. These people did not understand the reality that the economic model is itself failing - with or without the virus.

We spend hours in traffic, inhaling the polluted air and work long hours. Most people suffer from anxiety and/or stress, with a significant proportion needing psychological counselling, antidepressant drugs etc.

Covid-19 forced us into telecommuting and showed us that we can do more with less, and save time and start enjoying the simple things in life, like gardening, cooking, baking, and even walking.

But human slavishness to habit and pressures from the various businesses who cannot survive financially in a more efficient world, will gradually push some humans back to their usual ruts, like hamsters in their wheels.

While some young people, who will one day become old people, find that it is acceptable to sacrifice their grandparents and parents to Covid-19, many senior citizens themselves are also in their own ruts, their own hamster wheels.

Industrialisation has conditioned us to think that everyone must work, make money, then retire and enjoy the *golden* years.

Many people who retire often lose their purpose in life, become lazy and start accumulating illnesses, physical and mental. They often pay no heed to the *Use it or lose it* principle.

Employers often discriminate against older people.

People often suggest that retirees do voluntary work just to pass their time – I find this another form of exploitation – just like using unpaid young interns to get real work done.

The people who sell funeral insurance come knocking. Scammers target older people too.

Retired people who do nothing often have more time to see their doctors – some even fall into the trap of imagining illnesses or start worrying about simple ailments that will be hardly noticed if they were busy doing things. And, they have time. So, they see their doctors.

Doctors start sending them for various blood and other tests. Of course, as people get older things will deteriorate. Doctors of course try to do their best to help.

Medical statistics are used by doctors to say things like, *If you are over 50 (or 55, or 60) and belong to a certain ethnic group, you need to start taking these medications.*

What I am suggesting, and doing, is that older people must continue to work as long as they are able to and get paid for the work they do. Keeping active is the key to keeping healthy. Without a purpose, life becomes meaningless and people get sick physically and mentally.

It is their choice if they want to do volunteer work, but it should not be an exploitation where someone else makes the money.

Employers need to recognise the expertise and wisdom older people possess and be prepared to take advantage of it and pay for that. On the other hand, older people need to understand that the world is changing and there may be better ways of doing things and be willing to adapt.

We need to flush out the thinking:
You work hard, even long hours, and make money.
Save for retirement.
Retire and enjoy your golden years.

Replace it with this new thinking:
Everyone works as long as they are able to.
Nobody works long hours.
Whole life is golden years for all.
We look after those who are unable to work, young or old.

Anything is possible

Is it possible for countries like New Zealand to stay Covid-free? Is it possible to improve the quality of life of the people of New Zealand while staying Covid-free? Is it possible to have a world-leading economy and still stay Covid-free?

Why is it necessary to stay Covid-free, apart from saving the lives of the elderly and those with diabetes, obesity, asthma and other respiratory disorders, cardio-vascular disorders and immune deficiencies?

Are the lives of the elderly, the obese and other vulnerable citizens not worth protecting? Or, are we as a country, happy to sacrifice nanas, grandpas and our immune-suppressed relatives for the sake of the economy?

Can our nanas, grandpas and the others who some consider as economic burdens, be made productive?

Yes, it is all possible for the Team of 5

Million that quashed Covid-19.

Conclusion

As Covid-19 rages across the world, New Zealand, having successfully contained the virus, has a once-in-a-lifetime opportunity to use our number-eight-wire mentality and become the *All Blacks* of economy and quality of life.

We can achieve this by; investing in innovative high-tech industries aimed at making the country self-sufficient in essential supplies; developing smarter businesses; valuing and rewarding people, young and old; training and re-training people to keep up with the changes;

eliminating waste; reducing working hours; and, increasing healthier recreation activities.

Acknowledgements – Photos – In order of appearance

Cover photo - Image by Bungeeinternational from Pixabay

Waterfall 1 - Image by David Mark from Pixabay

Waterfall 2 - Image by Simon Steinberger from Pixabay

Chimney - Image by Jwvein from Pixabay

Aircraft - Image by WikiImages from Pixabay

Cruise ship - Image by Michelle Maria from Pixabay

Steam engine - Image by Wolfgang Eckert from Pixabay

Coronavirus - Image by Fernando Zhiminaicela from Pixabay

Face with mask - Image by Christo Anestev from Pixabay

Glacier - Image by Veronica Bosley from Pixabay

Nuclear Explosion - Image by WikiImages from Pixabay

Research - Image by Felixioncool from Pixabay

Masks - Image by leo2014 from Pixabay

Cows - Image by Waltteri Paulaharju from Pixabay

Queenstown - Image by Masa Hu from Pixabay

Hobbits - Image by Pawel Grzegorz from Pixabay

Jetboat - Image by Holger Detje from Pixabay

$ sign and virus - Image by Fernando Zhiminaicela from Pixabay

Lecture hall - Image by Nikolay Georgiev from Pixabay

Mask - Image by Tumisu from Pixabay

Whangarei - Image by 377053 from Pixabay

Learning - Image by Seila800 from Pixabay

Suicide - Image by Goran Horvat from Pixabay

Desperate - Image by Anemone123 from Pixabay

Exam - Image by Tiểu Bảo from Pixabay

Drone - Image by DJI-Agras from Pixabay

Robot - Image by Antonio Hernández from Pixabay

Traffic jam - Image by Alexander Grishin from Pixabay

Office - Image by David Mark from Pixabay

Surfing - Image by Free-Photos from Pixabay

Tai Chi - Image by Rene Rauschenberger from Pixabay

Having brought coronavirus under control, New Zealand has a once-in-a-lifetime opportunity to take advantage of the disruption caused by Covid-19 and become the All Blacks of economy and quality of life.

We can do this without wearing face masks, practising physical and social distancing and sacrificing the lives of nanas, grandpas and others who are easy pickings for the virus.

Innovative high-tech industries, much higher GDP, skilled workforce, smarter management, reduced waste, self-sufficiency, higher wages, reduced working hours, and healthier recreation, and a higher quality of life are all achievable goals.

Let's do it.

Comments and suggestions, most welcome.

ganeshan8@hotmail.com

@Ganeshan

www.ingramcontent.com/pod-product-compliance
Lightning Source LLC
Chambersburg PA
CBHW040327220526
45473CB00009B/2597